P9-CRB-896

AFRICAN ANIMALS

CAROLINE ARNOLD

Morrow Junior Books • New York

PHOTO CREDITS: Arthur P. Arnold, pages 6–7, 48; Caroline S. Arnold, pages 2–3, 12–13, 15, 17, 30–31, 37, 40–41, 45, 46–47; Carolyn T. Arnold, pages 4–5, 43; Eliot Brenowicz, pages 10–11, 32–33; Owen Floody, jacket and pages 1, 8–9, 20–21, 22–23, 24–25, 26–27, 34–35, 38–39; Susan Kean, pages 18–19; Wildlife Safari, Winston, Oregon, page 29.

The text type is 17-point Weiss. Book design by Christy Hale

Printed in Hong Kong by South China Printing Company (1988) Ltd. 1 2 3 4 5 6 7 8 9 10

Library of Congress Cataloging-in-Publication Data Arnold, Caroline. African animals/Caroline Arnold. p. cm. Summary: Describes animals of the African plains, forests, jungles, and deserts, and explains how each is able to adapt to its special environment. ISBN 0-688-14115-3 (trade)—ISBN 0-688-14116-1 (library) 1. Zoology—Africa—Juvenile literature. 2. Zoology—Africa—Pictorial works—Juvenile literature. [1. Zoology—Africa.] I. Title. QL336.A74 1997 591.96—dc20 96-16964 CIP AC

CONTENTS

A GIANT CONTINENT

AS THE MORNING SUN RISES, THE COOL AIR echoes with a lion's mighty roar. He has been hunting all night, but now he is ready for a morning nap. For animals all over Africa, it is time to start a new day.

Africa is the world's second-largest continent and is home to an amazing number of wild animals. They live on grasslands, in forests, and even in hot, dry deserts. Each place, or habitat, provides food and shelter for many kinds of animals. Let's discover how some African animals live and find the things they need to survive.

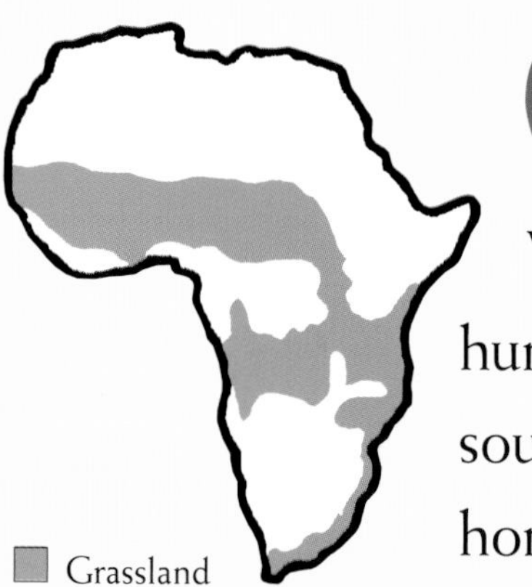

GRASSLANDS

VAST GRASSLANDS STRETCH FOR hundreds of miles across central, eastern, and southern Africa. These huge open spaces are home to an enormous variety of wildlife.

Millions of antelope, zebras, and other plant eaters roam the grasslands, searching for food and water. But they must watch out for lions, cheetahs, and other meat eaters that hunt them for food.

UGANDA KOB

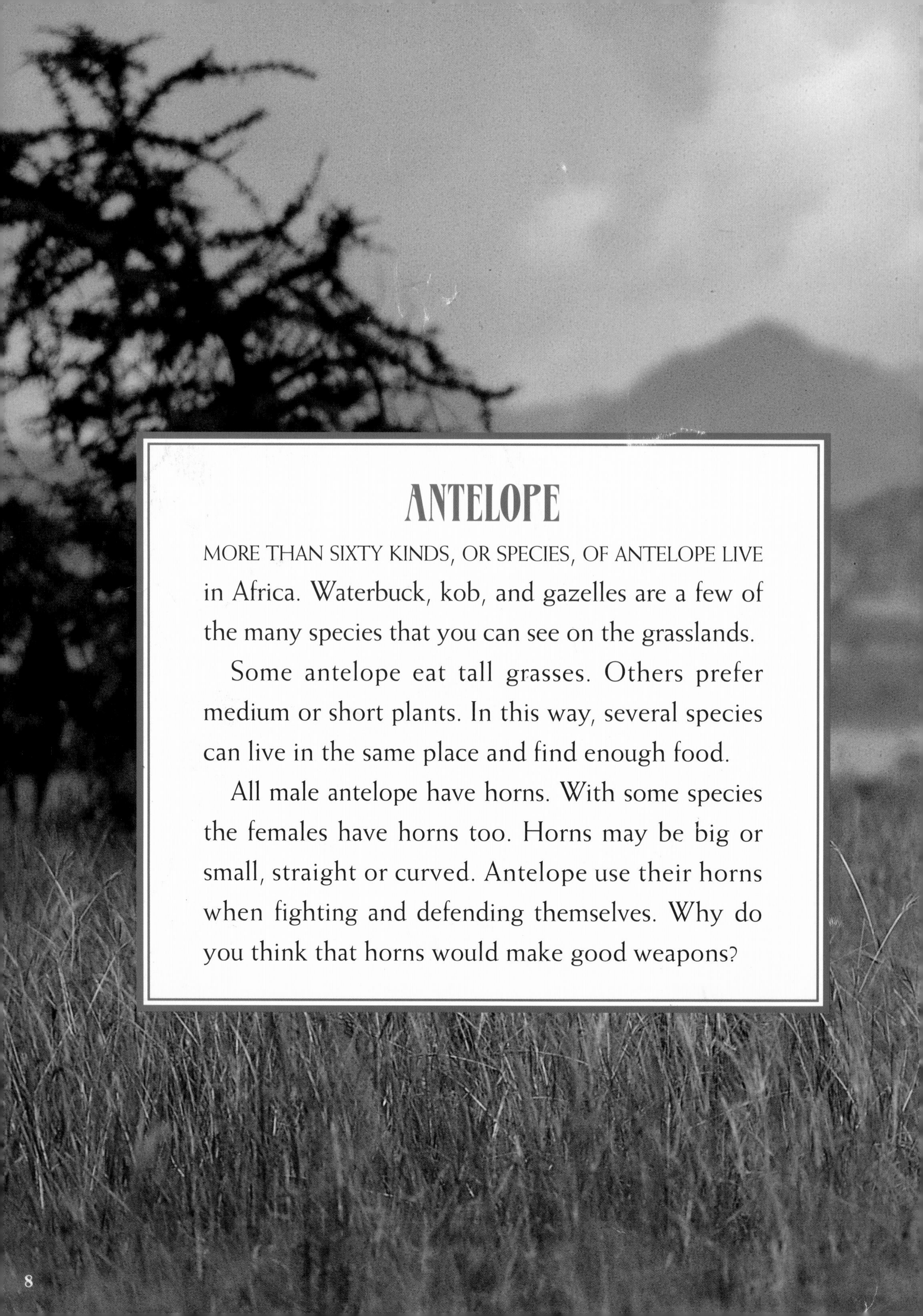

ANTELOPE

MORE THAN SIXTY KINDS, OR SPECIES, OF ANTELOPE LIVE in Africa. Waterbuck, kob, and gazelles are a few of the many species that you can see on the grasslands.

Some antelope eat tall grasses. Others prefer medium or short plants. In this way, several species can live in the same place and find enough food.

All male antelope have horns. With some species the females have horns too. Horns may be big or small, straight or curved. Antelope use their horns when fighting and defending themselves. Why do you think that horns would make good weapons?

WATERBUCK

ZEBRAS

ZEBRAS ARE THE HORSES OF THE AFRICAN PLAINS. LIKE OTHER horses, they have sturdy bodies and long, strong legs that are good for running.

Zebras help one another look and listen for danger. If one

ELEPHANTS

ELEPHANTS ARE THE HEAVIEST OF ALL LAND ANIMALS. Some adults weigh more than fourteen thousand pounds. That's as much as a medium-sized truck! Because elephants are so big, other animals cannot easily harm them.

One of the elephant's most unusual parts is its long hollow trunk. The elephant uses it for touching, breathing, smelling, and putting food into its mouth.

Sometimes an elephant sucks up water in its trunk. Then it squirts the water into its mouth for a drink or over its body for a shower.

Baby elephants must learn how to pick things up with their trunks. They get better with practice.

RHINOS

A RHINOCEROS IS ONE OF THE STRANGEST-LOOKING animals on earth. With its long horns and heavy body, a rhino looks a little like a living armored tank.

Rhino horns are made of the same hard material as your fingernails. Rhinos use their horns when fighting or defending themselves.

Sometimes rhinos dig holes with their horns. Scientists think they may do this to find salt and other minerals they need to eat.

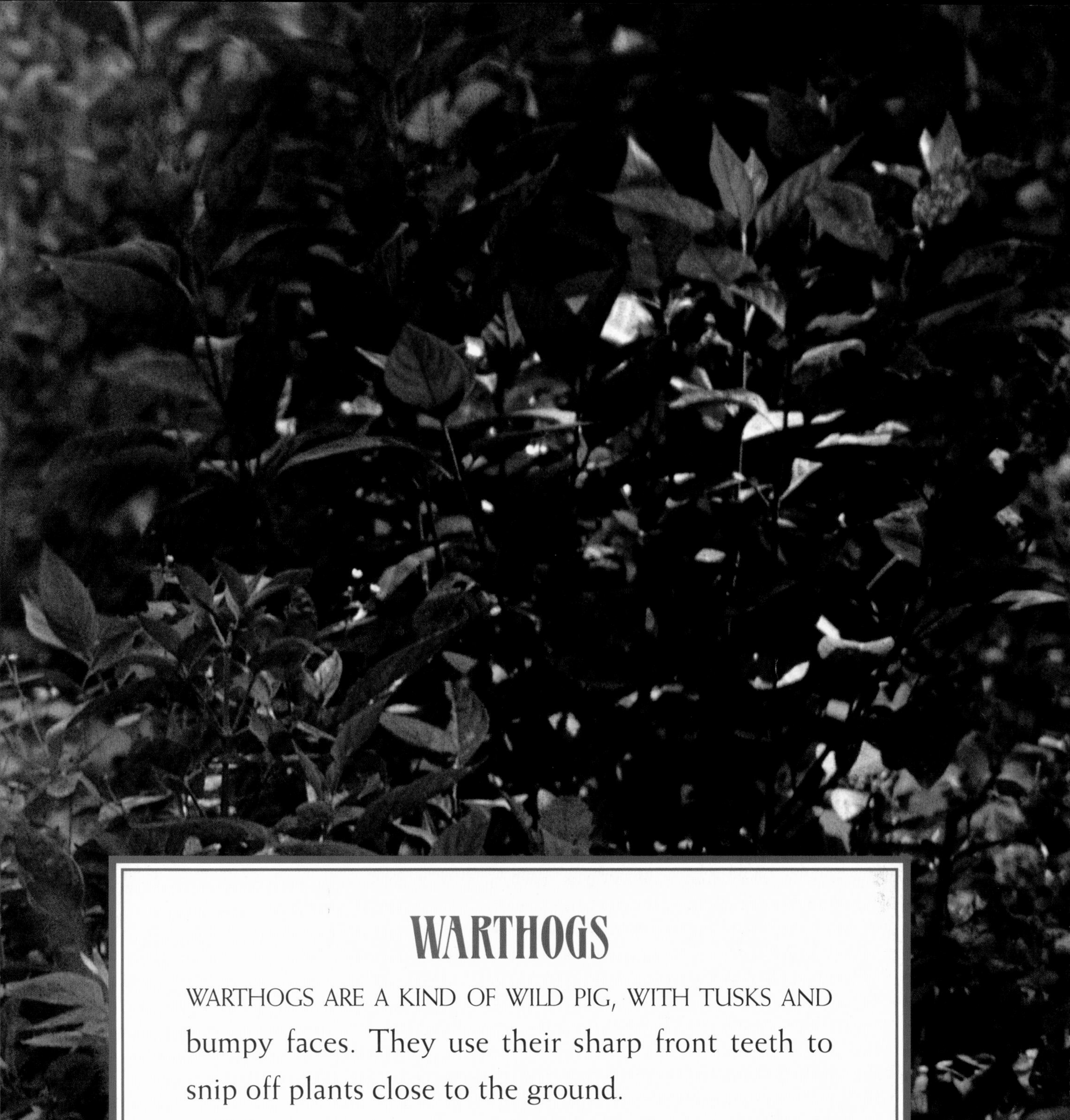

WARTHOGS

WARTHOGS ARE A KIND OF WILD PIG, WITH TUSKS AND bumpy faces. They use their sharp front teeth to snip off plants close to the ground.

When warthogs eat, they often kneel on their front legs to get closer to their food. But if something startles them, they jump up and trot away briskly.

OSTRICHES

WITH THEIR LONG LEGS AND FANCY FEATHERS, ostriches look something like giant ballerinas. Ostriches may grow to be eight feet tall and weigh as much as three hundred pounds. No other birds are as big as they are.

Ostriches are so heavy that they cannot fly. They run instead! They roam the African plains, looking for plants, insects, and small animals to eat.

During the mating season males sometimes fight over a female. They hiss and try to strike each other with their strong beaks.

HIPPOS

HERDS OF HIPPOPOTAMUSES LIVE IN MANY AFRICAN LAKES and rivers. When these huge animals are in the water, they look like small gray submarines.

Hippos spend most of each day in the water. Hippos can eat, walk, and even sleep underwater. They come to the surface every few minutes to get breaths of fresh air.

At night the hippos go onshore to eat grass and other plants.

CROCODILES

YOU CAN SEE CROCODILES ON THE BANKS OF MANY AFRICAN rivers and lakes. When thirsty animals go to drink, they must watch out for these huge reptiles.

Crocodiles often lie in the water with only their eyes and noses showing. Their rough skin makes it easy to

mistake them for bumpy logs floating in the water.

Crocodiles usually eat fish but sometimes hunt larger animals that step into the shallow water near the shore. Crocodiles grab their victims with their sharp teeth and pull them into deep water.

LIONS

NEARLY A DOZEN MEMBERS OF THE CAT FAMILY LIVE IN Africa. Good eyesight and hearing, quick reflexes, and sharp teeth and claws make all of the cats expert hunters. If you have a pet cat, you can watch it stalk and pounce in the same way that lions and other wild cats do.

Animals like lions that hunt for their food are called predators. The animals that they hunt are their prey. Lions are so strong that they can easily kill an animal the size of a zebra. A lion kills its prey by knocking it down and then biting it on the mouth and throat.

Lions live in groups called prides. Females in the pride do most of the hunting. They share their food with the males and the cubs.

LIONESS

CHEETAHS

CHEETAHS ARE THE SPRINTERS OF THE ANIMAL WORLD and chase their prey with amazing bursts of speed. Cheetahs can run as fast as seventy-five miles an hour. That's faster than any other land animal can go!

Strong claws on the cheetah's feet help it to grip the ground when it is running fast. Powerful back legs help it spring forward.

You can recognize a cheetah by the dark markings on either side of its nose.

MEERKATS

A MEERKAT IS A KIND OF MONGOOSE THAT LIVES ON THE African plains. It makes its home in an underground burrow.

Meerkats eat insects, lizards, and other small animals. When a group of meerkats goes out hunting for food, one of them acts as a sentry and looks out for larger predators that might be hunting *them.* If it spots danger, it calls out a warning, and all the meerkats race back to the burrow.

VULTURES, JACKALS, AND HYENAS

VULTURES, JACKALS, AND HYENAS ARE MEAT EATERS TOO, BUT they do not usually kill. They are scavengers, and they search for animals that are already dead. They also eat the leftover scraps of animals killed by lions, cheetahs, and other predators.

Vultures soar over the African plains and look for dead or dying animals on the ground. When they land, jackals, hyenas, and other scavengers often come to see if there is something good to eat.

Hyenas will steal food from other meat eaters if they get the chance.

VULTURES AND JACKAL

FORESTS

IN AFRICA FORESTS ARE found in the warm, tropical western and central regions, on the edges of the cooler highland plains, and in the mountains.

Insects, birds, snakes, squirrels, monkeys, apes, and many other kinds of animals find food and shelter in the forest. Tall trees provide safe places for the animals to make nests, and leafy branches shade them from the hot sun.

Some animals prefer the treetops, while others inhabit the lower branches or the forest floor. One tree may be home for many kinds of animals.

COLOBUS MONKEY

GORILLAS

GORILLAS LIVE BOTH IN LOWLAND FORESTS AND ON wooded mountain slopes. These primates are the largest of all apes and may weigh up to six hundred pounds. Some other primates that live in Africa are chimpanzees, monkeys, and baboons.

Gorillas are gentle animals that spend most of each day searching the forest for leaves and fruit. Although the males sometimes look and act fierce, they rarely attack other animals unless they are threatened.

When night falls, each gorilla makes itself a leafy nest. A gorilla usually sleeps on the ground or in the lower branches of a tree. How would you like to sleep in a tree at night?

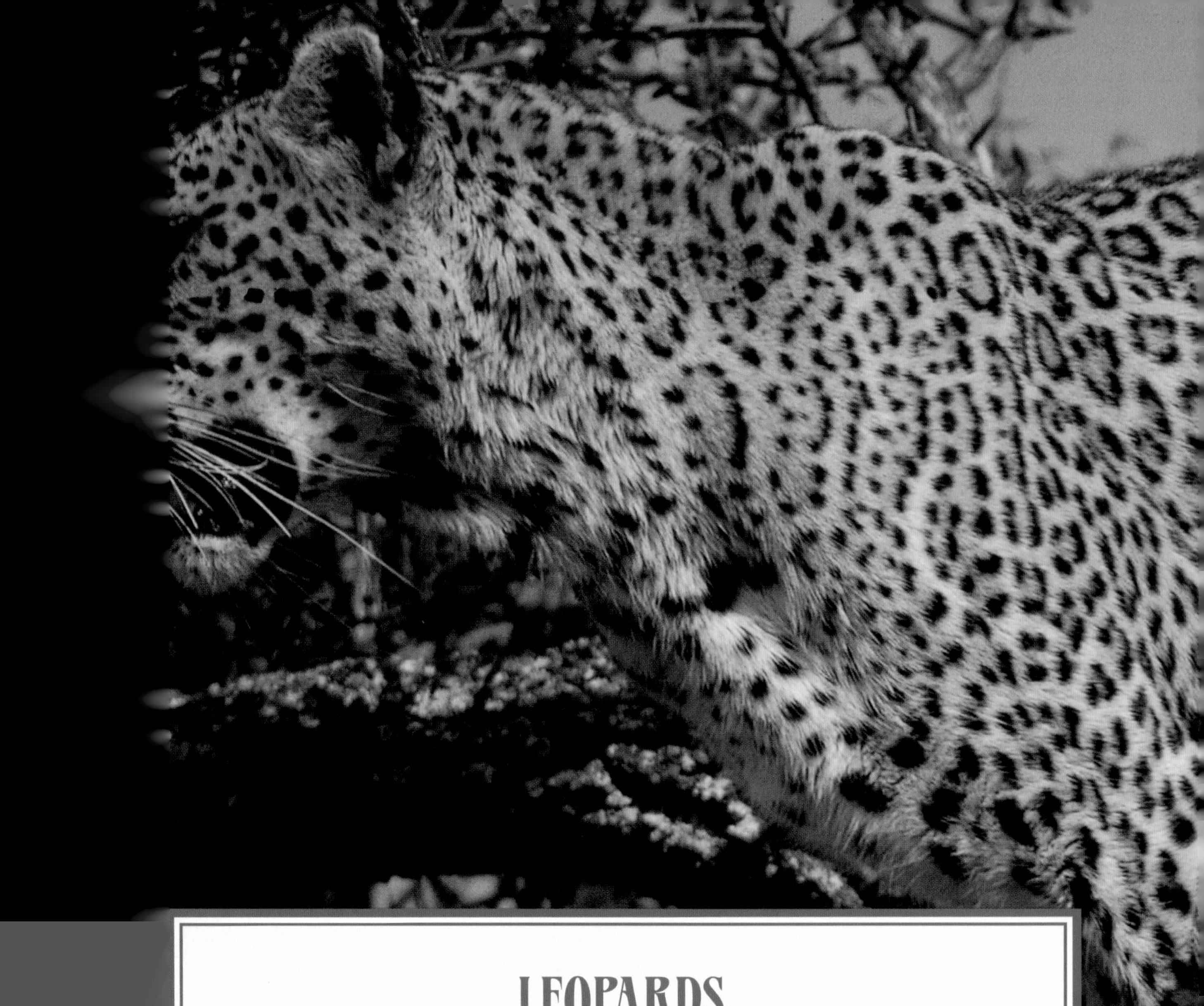

LEOPARDS

THE LEOPARD IS A POWERFUL LARGE CAT THAT LIVES IN woodlands at the edge of the African plains. This fearsome predator can see well in the dark and hunts mostly at night. It is an excellent climber and often leaps onto its prey from branches or rocks. It also hunts tree-dwelling animals like monkeys and birds.

When a leopard is stalking prey, its rosette-shaped spots help it hide in the shadows.

SNAKES

THE AFRICAN FOREST IS HOME TO MANY SPECIES OF snakes. They range in size from tiny vipers no thicker than your finger to pythons more than fifteen feet long.

Some snakes climb trees to hunt for birds and insects. They use their scaly bodies to grip the branches. Other snakes live on the ground, where they hunt rodents and other small animals. Because snakes have thin bodies, they can easily crawl into holes both to look for prey and to escape from animals that prey on them.

Desert
Semidesert

HUGE DESERTS STRETCH across northern Africa and much of the south. Bordering them are large semidesert regions that get a little more rain but are still without water much of the year.

One place in the desert where you can find water is an oasis. While some desert animals get enough water in their food, others go to an oasis to drink.

A few of the animals that make the desert their home are camels and some species of foxes, antelope, snakes, and lizards. They know how to survive in the desert's harsh climate.

CAMELS

A CAMEL IS WELL SUITED TO DESERT LIFE. ITS LARGE flat feet are good for walking in loose sand. And it can go for a week or more without drinking. When a camel does find water, it may drink as much as thirty-five gallons at once. That's enough water to fill a small bathtub!

The camel's big teeth are good for chewing the tough leaves of desert plants. When food is scarce, the camel gets energy from fat stored in its hump.

Although camels were once wild in the Sahara Desert, in northern Africa, almost all of them there now belong to people. The camels are used for riding and to carry things.

FOXES

THE TINY FENNEC FOX IS ONE OF SEVERAL KINDS OF foxes found in Africa. The light-colored coat of this desert dweller reflects the sun's rays, and fur protects the bottom of its feet from the hot ground. Body heat escapes through the fox's large ears, which helps keep the animal cool.

The fennec fox spends most of each day sleeping under a shady bush or in its underground den. At night it searches for insects, small reptiles, rodents, and plants to eat.

Foxes are members of the dog family. Does the fennec fox remind you of other dogs in some ways?

VANISHING WILDLIFE

MORE LARGE ANIMALS LIVE IN AFRICA THAN anywhere else on earth. However, as wild areas in Africa disappear, the number of animals grows smaller each year. By learning how African animals live at home in the wild, we can find out ways to help them and give them a better chance to survive in the future.